APP MAN S

PRESENTS...

NEW Edition

9 ROOKIE APP MISTAKES! OOPS!

...SO YOU CAN AVOID THEM!

CAN YOU BEAT THE APP MAN?
-TRY THE APPY-QUIZ

PHONE

SPECIAL INVITE

READ INSIDE...

INSIDER SECRETS • REVELATIONS • APPY-QUIZ •
APP-MAN TOP-TIPS • SPECIAL INVITATIONS • AND MUCH MORE...

APPY-QUIZ

CAN YOU BEAT THE APP MAN?

Q1: In 2011 over 300 MILLION Apps were sold, but in how long ?
- A: 1 month
- B: 2 weeks
- C: 1 day

Q2: Facebook the Social Media giant, bought which Mobile App for an impressive $1 BILLION?
- A: GoBiker
- B: Instagram
- C: YesMe

Q3: The Mobile Industry is bigger than?
- A: The book industry
- B: The music industry
- C: Both

Q4: When will mobile payments out-number over-the-counter cash payments?
- A: 2020
- B: 2040
- C: 2060

Q5: Who shops more via mobile phone Apps?
- A: Men
- B: Women
- C: No one knows

Q6: How much will mobile content advert sales be worth in 2012?
- A: 518 MILLION
- B: $1 BILLION
- C: $68 BILLION

Did you get them all right? Find out, find the answers at bottom left of this page

AppManSecrets.com

CONTACTS/CREDITS:

Web:
AppManSecrets.com

Email:
info@AppManSecrets.com

Tel:
+44 (0) 845 301 0042

App Advice:
The App Man

App Marketing:
The App Fairy

© Copyright 2012 ThinkEmotion Ltd

Q1:C Q2:B Q3:C Q4:A Q5:A Q6:C

Appy-Word-Search

A	N	A	Y	S	A	E	W	C	M	N	T
N	A	Y	M	F	H	E	M	O	C	N	I
D	N	P	E	I	L	Y	U	E	V	I	N
R	S	I	P	H	O	N	E	C	M	N	S
O	W	Y	M	M	H	D	Q	H	J	P	I
I	E	A	E	U	A	L	W	C	M	N	D
D	R	Y	M	F	H	N	Q	H	J	P	E
T	S	S	E	C	R	E	T	S	M	N	R
E	T	D	M	F	H	D	Q	H	J	P	I
S	N	A	E	U	Z	L	E	A	R	N	T
T	O	P	T	I	P	S	Q	H	J	P	I
S	N	I	E	P	P	A	H	C	I	R	T

ANDROID	EASY	IPAD	RICH APP
ANSWERS	INSIDER	INCOME	SECRETS
APP MAN	IPHONE	LEARN	TOP TIPS

ROOKIE MISTAKES #9

TRY TO WRITE THE CODE THEMSELVES

TO CREATE AN AMAZING BUILDING, YOU FIRST HAVE TO GET THE ARCHITECTURE RIGHT, YOU DON'T START BY TRYING TO LEARN HOW TO LAY THE BRICKS... IT'S THE SAME WITH APPS!

CONCENTRATE ON THE YOUR OBJECTIVES, THE APP GOALS, AND THE APP REVENUE MODEL. NOT THE CODE WRITING!

ROOKIE MISTAKES #5

...CUT CORNERS TO SAVE TIME & MONEY

PG-8

App Rookies can often be seduced into Cutting Corners in an attempt to get their App to market sharpish – DON'T DO IT at the Expense of Quality!

ROOKIE MISTAKES #4 — DON'T TEST THEIR CONCEPT!

APP MAN SECRETS – TOP TIP:

YOU'RE BEST ADVISED TO FOLLOW OUR APPROACH OF – 'TEST IT BEFORE YOU BUILD IT!"

I CAN SHOW YOU HOW, 1ST SIGN A NDA
(NON DISCLOSURE AGREEMENT)

www.bit.ly/IDFfjt

THEN REQUEST AN APP INSPECTION REPORT

www.bit.ly/GetAppReport

CASE SENSITIVE

"REMEMBER: TEST IT BEFORE YOU BUILD IT!"

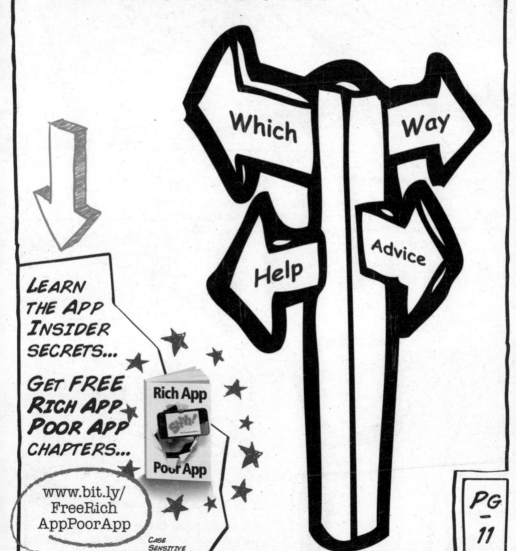

Product:		Price:	Quantity:	Total:
			Sub total:	
			TOTAL:	

Your Details:

Title _____ First name _____ Surname _____

Address: _____

_____ Postcode _____

Telephone number _____ Email _____

Your Payment: To purchase via website, visit www.AppManSecrets.com/products

Card type ☐ Visa ☐ Mastercard ☐ Maestro

Card number: ☐☐☐☐ ☐☐☐☐ ☐☐☐☐ ☐☐☐☐

Valid From ☐☐ / ☐☐   Expiry Date ☐☐ / ☐☐   Issue No. ☐

Currency: £GBP ☐ $ USD ☐   Total payment: _____

Card Holder's signature _____ Date _____

APPMANSECRETS.COM